W9-CCE-344

HOW IT WORKS

SMARTPHONES

by Lisa J. Amstutz

FOCUS READERS

WWW.FOCUSREADERS.COM

Focus Readers is distributed by North Star Editions:
sales@northstareditions.com | 888-417-0195

Produced for Focus Readers by Red Line Editorial.

Content Consultant: Dr. Sherali Zeadally, Associate Professor, College of Communication and Information, University of Kentucky

Photographs ©: DragonImages/iStockphoto, cover, 1; PeopleImages/iStockphoto, 4–5; AP Images, 7; Detchart Sukchit/Shutterstock Images, 8–9; Peter Howell/iStockphoto, 10; Andrey Popov/Shutterstock Images, 13; Ridofranz/iStockphoto, 14–15; beer5020/iStockphoto, 17; Zern Liew/iStockphoto, 18; milindri/iStockphoto, 20–21; Pieter Beens/Shutterstock Images, 22–23; hocus-focus/iStockphoto, 25; George Clerk/iStockphoto, 26–27; ersinkisacik/iStockphoto, 29

ISBN
978-1-63517-236-2 (hardcover)
978-1-63517-301-7 (paperback)
978-1-63517-431-1 (ebook pdf)
978-1-63517-366-6 (hosted ebook)

Library of Congress Control Number: 2017935881

Printed in the United States of America
Mankato, MN
June, 2017

ABOUT THE AUTHOR

Lisa J. Amstutz is the author of more than 60 children's books. Her work has appeared in a variety of magazines as well. She specializes in topics related to science, nature, and agriculture. Lisa and her family live on a hobby farm in rural Ohio.

TABLE OF CONTENTS

CHAPTER 1

AN ALL-IN-ONE DEVICE

A soccer coach snaps a team photo and shares it on social media. Then she texts the players their new schedule. She can do all this with just a smartphone.

Smartphones are cell phones with computers inside. Early cell phones used **analog** signals to copy sound vibrations. They could only send voice messages.

Smartphones can make calls, send messages, connect to the Internet, and more.

In the early 1990s, cell phones began using **digital** signals. This allowed the phones to send text as well as sound, making them the first smartphones.

Early smartphones had simple black-and-white graphics. Users chose options with a keypad or **stylus**. Later smartphones had cameras. They could send pictures and images, too.

In 2007, Apple released the iPhone. The iPhone combined the features of a phone, web browser, music player, and

CRITICAL THINKING

How did the change from analog to digital affect the development of smartphones?

Motorola introduced this smartphone in 1997.

camera. It also had a finger-operated touchscreen.

Smartphones continued to improve through the years. New **apps** allowed users to play games, surf the web, and much more.

CHAPTER 2

SENDING SIGNALS

Cell phones send messages on radio waves. To send a message, a cell phone must be near a cell tower. Each cell tower has a large antenna. It picks up the cell phone's signals. The antenna then sends these signals down to a base station. The base station routes the signals to their destination.

Antennas may be on objects such as towers, steeples, and flagpoles.

Tall buildings or mountains sometimes block signals sent out by an antenna.

Signals may be sent to a different base station to reach another cell phone. To reach a **landline**, signals are sent using a system of cables.

The area around each base station is called a cell. Each cell has six sides. The cells are laid out like a honeycomb. Cells can be different sizes. If there are many people in an area, such as in a city, the cells are small. If there are fewer people in an area, the cells are larger.

Whenever a cell phone is turned on, it is sending signals to the nearest cell tower. These signals allow the phone company to track where the phone is. The signals also tell the company how long the phone has been in use.

Each cell phone uses two different channels. One channel is for sending signals. The other is for receiving signals. That keeps the signals from interfering with each other.

Smartphones use a third channel as well. This channel carries data. It allows smartphones to access the Internet. This is what allows users to view websites and send e-mail. Many apps require Internet access, too. To use this channel, a smartphone needs access to broadband service. Smartphones can use this service to connect to the Internet anywhere they have a phone signal. But most cell phone companies have limits on how much data a phone can use.

Smartphones can also connect to the Internet using a Wi-Fi network. To do this, the phone must be close to a Wi-Fi modem or transmitter. The modem or

Data allows smartphone users to stream videos or play online games.

transmitter sends out a signal. If the phone has permission to connect to the network, it can pick up the signal and access the Internet. Homes, schools, and hotels often have Wi-Fi networks.

CHAPTER 3

SMARTPHONE PARTS

A smartphone's outer shell is called the casing. It holds all the phone's parts together. Most phones have a power button and volume control on the outside. Various ports are used for plugging in a charger, accessing files, or adding memory. Most smartphones also have a camera and a flash.

People use the touchscreen on a smartphone's front to operate the phone.

The smartphone's **hardware** is inside the casing. An antenna sends out the coded signals to the nearest cell tower. Smartphones get power from batteries. Most use a lithium-ion battery. This battery is small. But it is powerful and long lasting. It allows the phone to run for hours without being plugged in. It can be recharged, too.

The battery powers the phone's circuit board. The circuit board controls the phone's computer. It is made up of several smaller chips. One chip contains the microprocessor. This part manages information. It also works the keyboard and display. Other chips

Most phones have a charging cord that plugs into a USB port or an outlet.

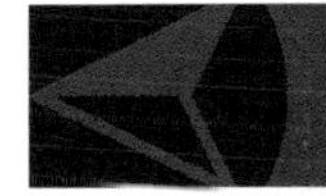

store information. These parts hold the computer's read-only memory (ROM), random access memory (RAM), and flash memory. ROM cannot be changed. It is used to start up the phone.

Flash memory can be changed. Like ROM, it stores information even when the phone is off. RAM is used to run **software**. The contents of this memory are lost when the phone is turned off.

PARTS OF A SMARTPHONE

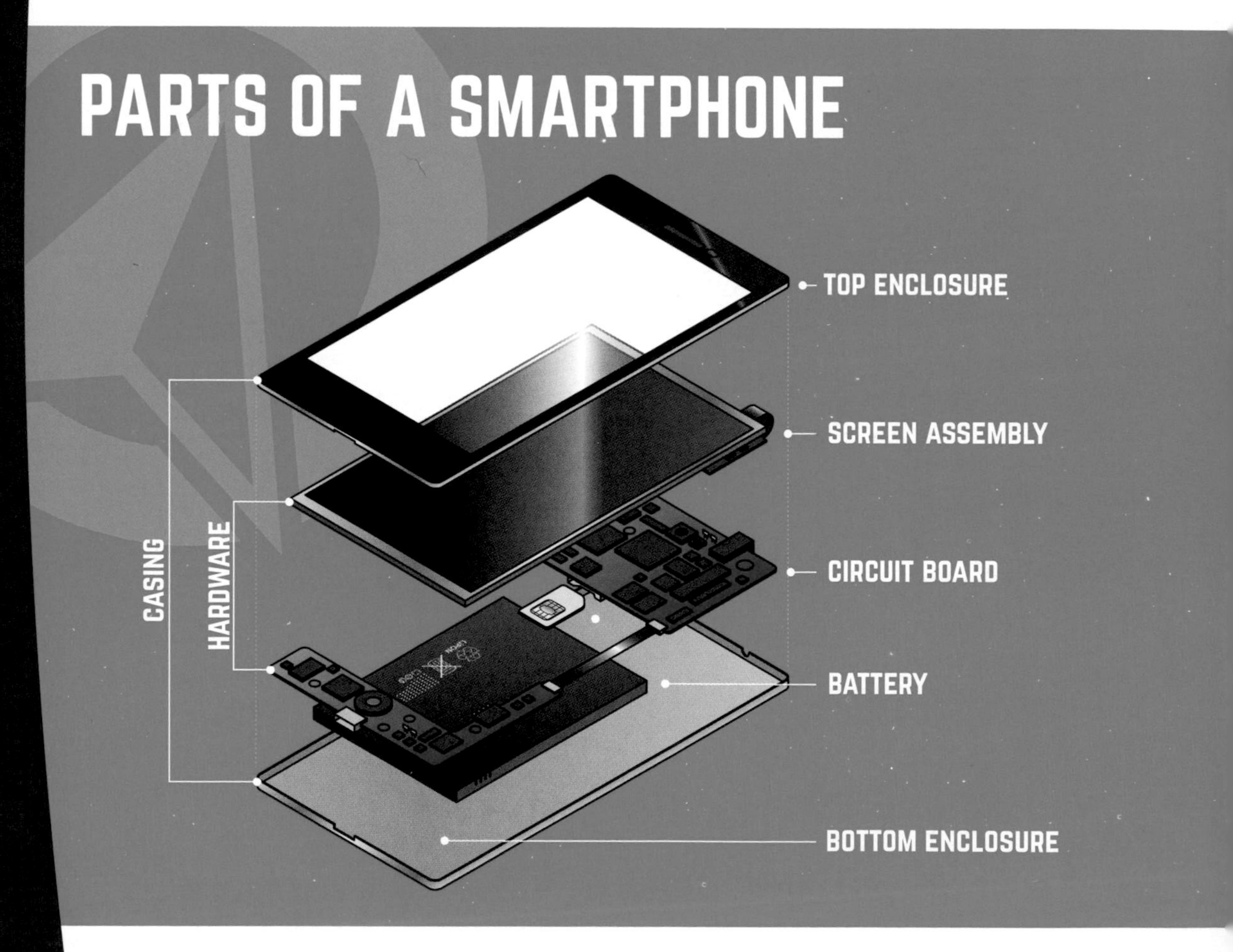

A microphone picks up the user's voice. When people speak, their vocal cords make the air around them vibrate. This generates a pattern of waves. The microphone captures this pattern and changes it into electrical signals. Then the microprocessor converts the signals into a series of zeroes and ones called binary code. Text and images can also be sent as binary code.

The coded signals are picked up by another phone. The microprocessor in that phone converts the signals back into sound waves. Then the phone's speaker puts out the sound.

TOUCHSCREEN

Most touchscreens respond to electrical currents in the user's finger. These screens have several layers. The bottom layer is a liquid crystal display (LCD). The LCD shows words or images. Above the LCD is a layer of glass. A layer of clear material sits above the glass. This material conducts electricity. A small amount of voltage from the battery gives this layer a slight charge. Another thin, clear material is on top.

When a person's finger touches the screen's top layer, the **conductive** layer's electrical charge changes. Sensors around the edge of the screen measure exactly where the screen was touched. They send a message to the microprocessor. Then the microprocessor responds to the command.

Touching the screen sends commands to the circuit board inside the phone.

14:59 PM
17
Photos
Calendar
Calculator
Music
Trash
Search
Maps
Chat
Notes
Scan
Preferences
Store
Files
Games
Videos
Mail
Settings
Home

SAMSUNG
21:10
5
Telephone
Contacts
Messages
Applications
9:41
Monday
10
Messages
Calendar
Weather
Clock
Notes
Reminders
Newsstand
iTunes Store
Compass
Settings
Phone
Mail
Safari

CHAPTER 4

TYING IT ALL TOGETHER

Each smartphone has an operating system (OS). The OS is the main software program that runs the phone. It connects the display, keyboard, and other parts of the phone's system. It also runs the phone's other software, including the apps. The OS makes it easier for users to interact with the phone's computer.

Samsung smartphones (left) use the Android OS, while iPhones (right) use Apple's iOS.

The OS translates computer code into easy-to-use graphics and text. Smartphones can use different types of operating systems. The Android OS and Apple's iOS are the two most common.

The OS controls the phone's apps as well. Users download these programs from app stores. Apps show up as icons on the phone's screen. Tapping an icon launches the app.

Smartphone users can choose from hundreds of thousands of apps. Some apps are designed to be practical. These apps include calendars, e-mail services, and weather updates. Other apps are just for fun. They allow users to play games or share pictures with friends. Many apps rely on the smartphone's ability to connect to the Internet. Popular apps include social media and maps. And new apps are being developed all the time.

78%
18:54
George IV Bridge
toward Chambers St
Then
Teviot Pl
Teviot Pl
Bristo Pl
Brighton St
Bristo Pl
38 min
Chambers St
George
Cowgate
RE-CENTER

CHAPTER 5

WHAT'S NEXT?

Smartphones have changed the world in many ways. Smartphone users can easily keep in touch with friends and family. Some smartphone users live in areas with no landlines. Their phones give them access to news, health information, education, and more.

Many drivers rely on smartphones for directions.

Despite all these benefits, smartphones pose some dangers. Many car accidents are caused by people using their phones while driving. Spending too much time on a smartphone can also interfere with grades, work, and friendships.

Smartphone technology is changing quickly. Some phones use near field communication (NFC). An NFC chip allows phones to share information when they are held close together. Smartphone

CRITICAL THINKING

Why might using a smartphone interfere with friendships or getting good grades?

People can use NFC to make purchases with their smartphones.

users can use this technology to share photos or play games. Future smartphones may not need to be near a cell tower. They may use satellites instead. Engineers continue to explore new possibilities for what smartphones can do.

FOCUS ON
SMARTPHONES

Write your answers on a separate piece of paper.

1. Write a paragraph explaining the way a smartphone sends and receives information.

2. Which smartphone apps do you think are the most useful? Why?

3. What part of a smartphone controls the phone's computer?

    A. the antenna
    B. the casing
    C. the circuit board

4. What would happen if a smartphone did not have a broadband or Wi-Fi connection?

    A. The smartphone could not access the Internet.
    B. The smartphone could not turn on.
    C. The smartphone could not receive text messages.

Answer key on page 32.

GLOSSARY

analog
Measuring or representing data by using a continuously changing signal instead of by using numbers.

apps
Computer programs that complete a task.

conductive
Made of material that electricity, heat, or sound can travel through.

digital
Measuring or representing data by using patterns of zeroes and ones known as binary code.

hardware
The mechanical and electronic parts that make up a device.

landline
A telephone connected to a network of cables on poles or underground.

software
The programs that run on a computer and perform certain functions.

stylus
A small stick used to write or touch buttons on a screen.

TO LEARN MORE

BOOKS

Enz, Tammy. *The Amazing Story of Cell Phone Technology.* Mankato, MN: Capstone Press, 2014.

Mangor, Jodie. *Inventing the Cell Phone.* North Mankato, MN: The Child's World, 2016.

Pipe, Jim. *You Wouldn't Want to Live Without Cell Phones!* New York: Scholastic, 2015.

NOTE TO EDUCATORS

Visit **www.focusreaders.com** to find lesson plans, activities, links, and other resources related to this title.

INDEX

Answer Key: 1. Answers will vary; **2.** Answers will vary; **3.** C; **4.** A